How You Can Save Water

by Dionna L. Mann

Children's Press®
An imprint of Scholastic Inc.

Content Consultants
American Geosciences Institute

Library of Congress Cataloging-in-Publication Data Available

978-1-338-83708-7 (library binding) | 978-1-338-83709-4 (paperback)

10 9 8 7 6 5 4 3 2 1   23 24 25 26 27

Printed in China 62
First edition, 2023

Series produced by Spooky Cheetah Press
Book prototype and logo design by Book&Look
Page design by Kathleen Petelinsek, The Design Lab

TABLE OF CONTENTS

Every Drop Counts!

Nothing can live without water. Plants, animals, and people need water to drink. Some plants and animals even live in water! Water helps us grow food, stay clean, and run machines that make items like clothes and books.

Most of Earth is covered by water. Most of it is salt water. Only some of Earth's water is fresh water, the kind we use in our daily lives. And it can run out. Luckily there are lots of easy ways you can save water.

Brushing your teeth with the tap on can use more than 1 gallon of water per minute!

CHAPTER 1

Use Less

One easy way to save water is to use less of it. A great place to start is in the bathroom. When you brush your teeth, turn off the tap while you scrub. Then turn it back on to swish and rinse. Do the same thing when you wash your hands.

It takes 16 cartons of milk from the school cafeteria to equal one gallon.

You can also save water when you wash your body. Taking a bath is a big water waster. It can take more than 35 gallons of water to fill a bathtub. A shower uses 2 to 5 gallons of water every minute. If you took a five-minute shower instead of a bath, you could save at least 10 gallons!

Waiting for the water to heat up? Catch the cold water in a bucket and use it to water plants!

More than half of all indoor water use happens in the bathroom!

Make one water glass your "glass for the day" instead of using a new one each time you are thirsty. Then you will not have to wash as many dishes!

To save water while washing dishes, fill two dish basins with water. Use one to wash the dishes and one to rinse. If you do not have any basins, first wet the dishes. Turn off the water while you scrub. Then turn the water on to rinse. That will use a lot less water.

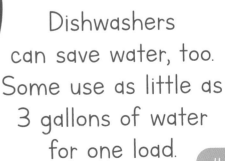

Dishwashers can save water, too. Some use as little as 3 gallons of water for one load.

CHAPTER 2

Make Small Changes

One drop of water leaking from a faucet may not seem like a lot. But one drop lost every second can waste a lot of water in a year. In fact, it would be enough water to fill more than 30 small kiddie pools! So fixing even the smallest of leaks can save water. Ask an adult to go on a "leak hunt"—and then help fix any you find!

Learning how to make small household repairs can be fun!

Some toilets save water by having two ways to flush. The button for liquid waste creates a half flush.

Some older toilets use 3 to 5 gallons for every flush. Talk to the adults in your house about switching to low-flow toilets. They use a lot less per flush. There is an even easier water-saving switch you can make. Ask an adult to put in showerheads that **conserve**, or save, water.

When you blow your nose, do not flush the tissues. Use a trash can instead.

Do you help with the laundry? Newer washing machines use less water than older machines. But both use a lot. No matter what type of machine you have at home, you should run it only when there is a full load of dirty clothes. The same goes for the dishwasher. Run it only when there is a full load.

All-natural laundry soaps help keep Earth's water clean.

Cactus plants do not need a lot of water.

You can also add mulch around your plants. That will help the soil hold in water.

18

CHAPTER 3

Make a Splash Outside

Do you have a garden at your home or at school that needs to be watered? Flower gardens are pretty. And they provide homes and food for helpful insects like bees. But some gardens use plants that need a lot of water. To save water, you can choose plants that do not need a lot of water. Consider plants that grow in your area naturally.

Gathering the water you need for your flower garden is another way to save water outside. You can do that by using a rain barrel. The rain barrel can catch rain as it falls from the sky and as it pours off a roof. This water is not safe for drinking, but you can use it to water the flowers or wash your family's car!

These kids are painting a rain barrel for their school.

Do you know how much water it takes to make a one-use plastic bottle that holds 8 ounces of water? It takes 256 ounces—32 times more water than the bottle holds.

A metal bottle keeps water cold—and can be used for years.

CHAPTER 4

Act Globally

Water is needed to make every human-made object. You can save water by reducing, reusing, and **recycling** the things you buy. An easy way to do that is by avoiding one-use plastic products, such as drinking straws and bottled water. Instead, use a reusable water bottle that you can fill up at home and carry with you.

Arrange a trip with your school or a club to visit some local waterways in your watershed. A watershed is a big area of land where water collects and runs into rivers, lakes, and streams nearby. Every community is part of a watershed. Learning about your local watershed can help you understand and conserve water as a **natural resource**.

These kids are visiting a river in their community's watershed.

SAVE WATER

We can all make a difference by learning about how to save water!

We can all save water. We can turn off the tap whenever we can. We can reduce, reuse, and recycle the things we buy. And we can speak up if we see water being wasted. Remember: Every drop counts!

Make a list of what you think are the top 10 ways you use water in your home. Talk to your family about each item on your list. How can you all work together to save water?

BE A LEAK DETECTIVE

A leaky faucet is easy to spot. But catching a leaky toilet can be tricky. Toilets can leak on the inside—from the fill tank into the bowl—so you cannot see the water dripping. A toilet leak can go on for a long time and waste hundreds of gallons of water. Here is a simple test to see if your toilet is leaking from the tank into the bowl.

YOU WILL NEED

- An adult's help
- Measuring cup
- Purple grape juice
- Timer

STEPS

1. Ask an adult to lift the toilet tank lid for you.
2. Pour in about ¼ cup of grape juice.
3. Replace the tank top.
4. Wait 20 to 30 minutes. Do not flush.
5. Look inside the toilet bowl.
6. If you see purple inside the bowl, then you have a leak.

WATER WARRIOR

MEET MILO CRESS

More than 500 million straws are used every day around the world. Making plastic straws wastes water. And used straws **pollute** our water. That inspired Milo Cress to start a campaign called Be Straw Free.

Thanks to Milo's efforts, restaurants started asking customers if they wanted a straw before giving one. Cities and states declared Skip the Straw days. Today, Milo challenges all kids to do their part. It can be as simple as saying "no, thanks" next time someone offers you a plastic straw.

GLOSSARY

conserve (kuhn-SURV) to save something from loss, decay, or waste

natural resource (NACH-ur-uhl REE-sors) something that is found in nature and is valuable to humans, such as water or trees

pollute (puh-LOOT) to add something that can harm living things

recycling (ree-SYE-kling) processing old items such as glass, plastic, newspapers, and more so they can be used to make new products

INDEX

ABOUT THE AUTHOR

Dionna L. Mann is a children's book author. Nothing delights her taste buds more than a glass of cold water on a hot day. And nothing delights her ears more than the calming roar of the ocean. You can find Dionna online at dionnalmann.com.